CGEIT®
REVIEW QUESTIONS, ANSWERS & EXPLANATIONS MANUAL 2015 SUPPLEMENT

CGEIT®
Certified in the
Governance of
Enterprise IT®
An ISACA® Certification

ISACA®

With more than 115,000 constituents in 180 countries, ISACA *(www.isaca.org)* helps business and IT leaders build trust in, and value from, information and information systems. Established in 1969, ISACA is the trusted source of knowledge, standards, networking, and career development for information systems audit, assurance, security, risk, privacy and governance professionals. ISACA offers the Cybersecurity Nexus™, a comprehensive set of resources for cybersecurity professionals, and COBIT®, a business framework that helps enterprises govern and manage their information and technology. ISACA also advances and validates business-critical skills and knowledge through the globally respected Certified Information Systems Auditor® (CISA®), Certified Information Security Manager® (CISM®), Certified in the Governance of Enterprise IT® (CGEIT®) and Certified in Risk and Information Systems Control™ (CRISC™) credentials. The association has more than 200 chapters worldwide.

Disclaimer

ISACA has produced the *CGEIT® Review Questions, Answers & Explanations Manual 2015 Supplement* primarily as an educational resource to assist individuals preparing to take the CGEIT certification exam. It was produced independently from the CGEIT exam and the CGEIT Certification Committee, which has had no responsibility for its content. Copies of past exams are not released to the public and were not made available to ISACA for preparation of this publication. ISACA makes no representations or warranties whatsoever with regard to these or other ISACA publications assuring candidates' passage of the CGEIT exam.

Reservation of Rights

ISACA

3701 Algonquin Road, Suite 1010
Rolling Meadows, Illinois 60008 USA
Phone: +1.847.253.1545
Fax: +1.847.253.1443
Email: *info@isaca.org*
Web site: *www.isaca.org*

Participate in the ISACA Knowledge Center: *www.isaca.org/knowledge-center*
Follow ISACA on Twitter: *https://twitter.com/ISACANews*
Join ISACA on LinkedIn: ISACA (Official), *http://linkd.in/ISACAOfficial*
Like ISACA on Facebook: *www.facebook.com/ISACAHQ*

ISBN 978-1-60420-531-2
CGEIT® Review Questions, Answers & Explanations Manual 2015 Supplement
Printed in the United States of America

PREFACE

ISACA is pleased to offer this *CGEIT® Review Questions, Answers & Explanations Manual 2015 Supplement*. The purpose of this manual is to provide the CGEIT candidate with sample questions and testing topics to help prepare and study for the CGEIT exam.

The material in this manual consists of 60 multiple-choice study questions, answers and explanations, which are organized according to the CGEIT job practice domains. These questions, answers and explanations are intended to introduce the CGEIT candidate to the types of questions that appear on the CGEIT exam. They are not actual questions from the exam. Questions are sorted by CGEIT job practice domains, and a sample exam of 60 questions is also provided. Sample questions contained in this manual are provided to assist the CGEIT candidate in understanding the material in the *CGEIT® Review Manual 2015* and to depict the type of question format typically found on the CGEIT exam. The CGEIT candidate may also want to obtain a copy of the *CGEIT® Review Questions, Answers & Explanations Manual 2015*, which consists of 180 multiple-choice study questions, answers and explanations.

ISACA wishes you success with the CGEIT exam. Your commitment to pursuing the leading certification for IT governance practitioners is exemplary, and ISACA welcomes your comments and suggestions on the use and coverage of this manual. Once you have completed your exam, please take a moment to complete the online evaluation that corresponds to this publication (*www.isaca.org/studyaidsevaluation*). Your observations will be invaluable as new questions, answers and explanations are prepared.

ACKNOWLEDGMENTS

This *CGEIT® Review Questions, Answers & Explanations Manual 2015 Supplement* is the result of the collective efforts of many volunteers. ISACA members from throughout the global IT governance profession participated, generously offering their talents and expertise. This international team exhibited a spirit and selflessness that has become the hallmark of contributors to this valuable manual. Their participation and insight are truly appreciated.

CGEIT Quality Assurance Team
Rashid Jamil, CISA, CISM, CGEIT, CRISC, CISSP, ITIL, PMP, Sysforte Corporation, USA
Felix R. Ramirez, CISA, CGEIT, CRISC, Riebeeck Stevens, Ltd., USA
Nancy J. Thompson, CISA, CISM, CGEIT, PMP, USA
Daniel van den Hove, Alphega, Belgium
Jurgen Van de Sompel, CGEIT, Effect-IT, Belgium

TABLE OF CONTENTS

Page intentionally left blank

INTRODUCTION

OVERVIEW

This manual consists of 60 multiple-choice questions, answers and explanations (numbered GS1-1, GS1-2, etc.). These questions are selected and provided in two formats.

Questions Sorted by Domain

Questions, answers and explanations are provided (sorted) by the CGEIT job practice domains. This allows the CGEIT candidate to refer to specific questions to evaluate comprehension of the topics covered within each domain. These questions are representative of CGEIT questions, although they are not actual exam items. They are provided to assist the CGEIT candidate in understanding the material in the *CGEIT® Review Manual 2015* and to depict the type of question format typically found on the CGEIT exam. The numbers of questions, answers and explanations provided in the five domain chapters in this manual provide the CGEIT candidate with a maximum number of study questions.

Sample Exam

A random sample exam of 60 questions is also provided in this manual. **This exam is organized according to the domain percentages specified in the CGEIT job practice and used on the CGEIT exam:**

Framework for the Governance of Enterprise IT .... 25 percent
Strategic Management .... 20 percent
Benefits Realization.. 16 percent
Risk Optimization .... 24 percent
Resource Optimization .... 15 percent

Candidates are urged to use this sample exam and the answer sheets provided to simulate an actual exam. Many candidates use this exam as a pretest to determine strengths or weaknesses, or as a final exam. Sample exam answer sheets have been provided for both uses. In addition, a sample exam answer/reference key is included. These sample exam questions have been cross-referenced to the questions, answers and explanations by domain so that it is convenient to refer to the explanations of the correct answers. This publication is ideal to use in conjunction with the *CGEIT® Review Manual 2015* and the *CGEIT® Review Questions, Answers & Explanations Manual 2015.*

It should be noted that the *CGEIT® Review Questions, Answers & Explanations Manual 2015 Supplement* has been developed to assist CGEIT candidates in studying and preparing for the CGEIT exam. While using this manual to prepare for the exam, please note that the exam covers a broad spectrum of governance of enterprise IT issues. Do not assume that reading and working the questions in this manual will fully prepare you for the exam. Because exam questions often relate to practical experience, CGEIT candidates are advised to refer to their own experience and to other publications referred to in the *CGEIT® Review Manual 2015.* These additional references are excellent sources of further detailed information and clarification. It is recommended that candidates evaluate the job practice domains in which they feel weak or require a further understanding and then study accordingly.

Please note that this publication has been written using standard American English.

TYPES OF QUESTIONS ON THE CGEIT EXAM

CGEIT exam questions are developed with the intent of measuring and testing practical knowledge and applying the governance of enterprise IT principles, practices and standards. As previously mentioned, all questions are presented in a multiple-choice format and are designed for one best answer.

Candidates are cautioned to read each question carefully. Many times, a CGEIT exam question will require candidates to choose the appropriate answer that is **MOST** likely or **BEST**. Other times, candidates may be asked to choose a practice or procedure that would be performed **FIRST** related to the other choices. In every case, candidates are required to read the question carefully, eliminate known wrong choices and then make the best choice possible. Knowing that these types of questions are asked on the exam and how to study to answer them will assist CGEIT candidates in successfully preparing for the CGEIT exam.

Each CGEIT question has a stem (question) and four choices (answers). Candidates are asked to choose the correct or best answer from the choices. The stem may be in the form of a question or an incomplete statement. All questions are presented in a multiple-choice format and are designed for one best answer.

Another condition that CGEIT candidates should consider when preparing for the exam is to recognize that the governance of enterprise IT is global and that individual perceptions and experiences may not reflect the more global position or circumstance. Because the CGEIT exam and manuals are written for the international community, candidates will be required to be somewhat flexible when reading a condition that may be contrary to their experience. It should be noted that actual CGEIT exam questions are written by experienced IT practitioners from around the world. Each question on the actual CGEIT exam is reviewed by ISACA's CGEIT Test Enhancement Subcommittee and CGEIT Certification Committee, both of which consist of international members. This manual has been reviewed by an international quality assurance team (QAT) specially put together to review the questions, answers and explanations.

Note: ISACA review manuals are living documents. As technology advances, ISACA manuals will be updated to reflect such advances. Further updates to this document before the date of the exam may be viewed at *www.isaca.org/studyaidupdates*.

Any suggestions to enhance the materials covered herein, or reference materials, should be submitted to *studymaterials@isaca.org*.

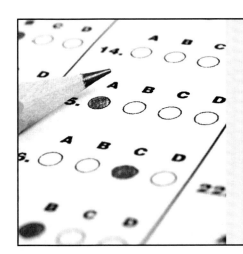

PRETEST

If you wish to take a pretest to determine strengths and weaknesses, the Sample Exam begins on page 35 and the pretest answer sheet is on page 47.
You can score your pretest with the Sample Exam Answer and Reference Key on page 45.

Page intentionally left blank

QUESTIONS, ANSWERS AND EXPLANATIONS BY DOMAIN

DOMAIN 1—FRAMEWORK FOR THE GOVERNANCE OF ENTERPRISE IT (25%)

GS1-1 Which of the following choices is the **MAIN** advantage of implementing a governance of enterprise IT framework?

    A.   Establishing and monitoring accountability for IT-related initiatives
    B.   Reducing IT-related risk by increasing IT investment
    C.   Reducing IT-related costs by achieving IT process improvements
    D.   Centralizing IT control through an IT steering committee

A is the correct answer.

Justification:
    A.   After the business strategy is defined and the business direction is clear, establishing and monitoring accountabilities for various IT-related initiatives is critical. This can be achieved by having in place a governance of enterprise IT framework.
    B.   Reducing IT-related risk is just one element of establishing and monitoring accountability.
    C.   Reducing IT-related cost is just one element of establishing and monitoring accountability.
    D.   Centralization of IT control through an IT steering committee may be beneficial; however, it is not the main objective of a governance of enterprise IT framework implementation.

GS1-2 When a new IT governance policy has been approved, it is **BEST** to:

    A.   have an independent party sign off.
    B.   conduct a walk-through exercise.
    C.   prepare a communication plan.
    D.   update the IT strategy accordingly.

C is the correct answer.

Justification:
    A.   Sign-off by an external party, when appropriate, should occur prior to approval.
    B.   Conducting a walk-through based on the revised policy should occur prior to approval.
    C.   When a document, such as a policy, has been updated, it is good practice to communicate those changes throughout the organization.
    D.   Relevant policy changes should be incorporated into the IT strategy, which should be included in the communication plan.

GS1-3 Which of the following choices is the **PRIMARY** reason for defining and managing the enterprise IT strategy?

    A.   It has become an industry standard.
    B.   It directs short-term IT goals.
    C.   It improves the efficiency of IT services.
    D.   It contributes to business value.

D is the correct answer.

Justification:
    A.   IT is part of the business strategy.
    B.   Short-term IT goals will be defined by the long-term goals in the IT strategy.
    C.   Improving efficiency of IT services is part of the IT strategy execution.
    D.   The enterprise IT strategy must be aligned with business objectives, which focus on value delivery to stakeholders.

GS1-4   Information security governance awareness is **BEST** established when:

    A.   senior management is supportive.
    B.   data ownership is identified.
    C.   assets to be protected are identified.
    D.   security certifications are issued.

A is the correct answer.

Justification:
    A.   **The best way to increase awareness in the enterprise is through guaranteed senior management championship.**
    B.   Data ownership identification is a necessary, but not sufficient, component to ensure information security governance awareness.
    C.   Assets protection is an operational mechanism to support data ownership.
    D.   Security certification is an operational mechanism to support asset protection.

GS1-5   A consulting firm re-engineered a customer trading system of an investment bank. Then the investment bank requested a security review of this system from the same consulting firm. From an IT governance perspective, which of the following choices is the **BEST** to consider?

    A.   Ensure that sensitive customer data are securely kept inside the consulting firm.
    B.   Ensure that a security assurance review plan is in line with regulatory requirements.
    C.   Ensure that segregation of duties (SoD) is in place within the consulting firm.
    D.   Ensure that the service level meets the criteria in the vendor due diligence policy.

C is the correct answer.

Justification:
    A.   As a consulting firm, sensitive information needs to be kept securely. As long as sensitive information does not leave the consulting firm, no serious consequences are envisioned. However, this is not relevant to the problem of conflict of interest.
    B.   A security assurance review plan must be in line with regulatory requirements, but is a secondary requirement to the segregation of duties (SoD) within the consulting firm.
    C.   **Careful consideration is required when a single vendor performs both implementation and its review. Independence needs to be secured when a review is made. When the same consulting firm conducts both implementation and its review, SoD may need to be checked in order to maintain the validity of review results.**
    D.   It is a fundamental requirement that the service level be compliant with the vendor due diligence policy of a sourcing organization. However, this is not relevant to the problem of conflict of interest.

GS1-6 Which of the following benefits is the **MOST** important for senior management to understand the value of governance of enterprise IT? It allows senior management to:

A. understand how the IT department works.
B. make key IT-related decisions.
C. optimize IT resource utilization.
D. evaluate business continuity provisions.

B is the correct answer.

Justification:
A. Understanding how the IT department works is the responsibility of the chief information officer (CIO). However, senior management should understand the role of IT.
B. When senior management understands the benefits of governance of enterprise IT as well as new technologies and challenges, they act as informed decision makers and take ownership of IT-related decisions.
C. Optimizing IT resource utilization is a subset of how the IT department works and is, therefore, the responsibility of the CIO.
D. Evaluating business continuity provisions is an operational responsibility of senior management.

GS1-7 Which of the following activities is the **MOST** essential for ensuring resource optimization within governance of enterprise IT?

A. Providing direction for strategic resources
B. Defining guidelines for performance indicators
C. Evaluating resource strategy against enterprise requirements
D. Establishing principles for management of resources

D is the correct answer.

Justification:
A. Providing direction is only part of the optimization process.
B. Providing guidelines for performance indicators is an operational activity that can be done based on the principles.
C. Evaluating resource strategy against enterprise requirements is dependent on establishing principles for management of resources.
D. Establishing principles for management of resources creates the framework for enabling allocation of optimized resources. ISACA's COBIT 5 framework states, "Define the principles for guiding the allocation of management of resources and capabilities so that IT can meet the needs of the enterprise, with the required capability and capacity according to the agreed-on priorities and budgetary constraints."

GS1-8   Which of the following choices has the **GREATEST** impact on the selection of an IT governance framework?

    A.   Corporate culture
    B.   Data regulatory requirements
    C.   Skills and competencies
    D.   Current process maturity level

A is the correct answer.

Justification:
    A.   **Corporate culture is the way that enterprises make decisions. Enterprises consider human factors, decision-making style, risk appetite, etc., and this has the greatest impact on the selection of an IT governance framework.**
    B.   Data regulatory requirements are important, but do not have the greatest impact when compared with corporate culture.
    C.   Skills and competencies are important, but these can be acquired at any time.
    D.   Current process maturity level is useful and important, but not as important as corporate culture.

GS1-9   When implementing governance of enterprise IT, which of the following factors is the **MOST** critical for the success of the implementation?

    A.   Improving IT knowledge of the board of directors
    B.   Decision making on IT investments by the board of directors
    C.   Documenting the IT strategy
    D.   Identifying the enablers and establishing performance measures

D is the correct answer.

Justification:
    A.   IT knowledge may be helpful, but it does not affect the outcome of the implementation of governance of enterprise IT.
    B.   Decision making on IT investments by the board of directors uses the IT governance framework.
    C.   The IT governance framework is used to document the IT strategy.
    D.   **Implementation of governance of enterprise IT includes identification of the enablers and the measurement of the goals.**

GS1-10   While implementing IT governance within an enterprise, the **PRIMARY** focus must be on the objectives of:

    A.   the enterprise.
    B.   the stakeholders.
    C.   the business function.
    D.   IT management.

B is the correct answer.

Justification:
    A.   Enterprise objectives are driven by stakeholder objectives.
    B.   **Enterprises exist to create value for their stakeholders.**
    C.   Business function objectives are outcomes of enterprise objectives.
    D.   IT management's objectives are derived from enterprise and business function objectives.

GS1-11 The **PRIMARY** focus in effective organizational change enablement of a governance of enterprise IT implementation should be on:

    A.   documenting the what and how of the change.
    B.   clarifying the reason to change.
    C.   communication of the vision.
    D.   demonstrating achieved results.

B is the correct answer.

Justification:
    A.   The what and the how of the change describes the impact of the change to the organization. However, the what and the how may not be effective if the why is not understood.
    B.   **The first action should be to work on the motivation of people by explaining the reasons why the change is necessary.**
    C.   Communication of the vision facilitates understanding of the change in roles and responsibilities for individuals. This can only be done when the reason for change has been defined.
    D.   Demonstrating achieved results to the organization shows the enterprise's ability to run change programs effectively. This is a normal activity as the enabled change progresses toward its objectives.

GS1-12 Which of the following choices is the **MOST** relevant in the enterprise culture change of an IT governance implementation?

    A.   Having employees who have values and beliefs
    B.   Having corporate values aligned with leaders in the industry
    C.   Having leaders who inspire new values
    D.   Having clearly communicated values and beliefs

C is the correct answer.

Justification:
    A.   Employees' values and beliefs may not be aligned with the enterprise's culture change.
    B.   Having corporate values aligned with leaders in the industry does not necessarily provide for a competitive advantage.
    C.   **The culture of an enterprise is a reflection of leadership consciousness—a reflection of the values, beliefs and behaviors of the leaders and the legacy of the past leaders—but enabling enterprise culture change will be more effective because inspiring leaders can make the organization align with their values.**
    D.   Having values and beliefs communicated does not mean they are understood and adopted.

GS1-13   Which of the following choices **BEST** describes the role of the board of directors in IT risk governance?

A.   Ensure that the planning, budgeting and performance of IT risk controls are appropriate.
B.   Assess and incorporate the results of the IT risk management activity into the decision-making process.
C.   Ensure that the enterprise risk appetite and tolerance are understood and communicated.
D.   Identify, evaluate and minimize risk to IT systems that support the mission of the enterprise.

C is the correct answer.

Justification:
A.   Ensuring that planning, budgeting and performance of IT risk controls are appropriate is an operational responsibility of the chief information officer (CIO).
B.   Assessing and incorporating the results of the risk management activity into the decision-making process is the responsibility of risk management.
C.   **The board of directors is responsible for setting the direction and boundaries for risk to be taken by the enterprise. Therefore, they need to understand and communicate the level of risk appetite and tolerance that they are ready to accept to effectively manage their business.**
D.   Identifying, evaluating and minimizing risk to IT systems that support the mission of the enterprise is an operational responsibility of the CIO.

GS1-14   Which of the following choices demonstrates the **GREATEST** influence of an IT governance framework for IT-related issues?

A.   To settle differences of opinion among board of directors members
B.   To resolve cross-departmental conflicts
C.   To gain loyalty from key stakeholders
D.   To investigate weaknesses within processes

B is the correct answer.

Justification:
A.   An IT governance framework is not designed to resolve disagreements among board of directors members.
B.   **An IT governance framework can exert its greatest influence in resolving cross-departmental conflicts for IT-related issues. When a governance framework is in place, business units are aligned to strategies and resource prioritization is made accordingly.**
C.   When good IT governance is in place, the enterprise may function more effectively, which may lead to increased loyalty from stakeholders.
D.   Risk management will identify weaknesses within processes.

GS1-15   Which of the following choices would **PRIMARILY** create transparency in the IT decision-making process as part of governance of enterprise IT?

A.   Stakeholder-approved roles, responsibilities, goals and metrics are communicated.
B.   The progress reporting process of service delivery is clearly established.
C.   Communication of decisions to IT employees is clear.
D.   Balanced scorecard (BSC) results are promptly communicated to the enterprise.

A is the correct answer.

Justification:
A.   **Transparency would be created by communicating stakeholder-approved roles, responsibilities, goals and metrics to the enterprise. This allows everyone to understand the basis for the decisions being made.**
B.   Progress reporting reflects the advancement of the execution, but does not provide information about the basis of the decision.
C.   Communication of decisions to employees is one piece of the process to create transparency in IT decision making.
D.   The balanced scorecard (BSC) provides information about how IT is performing according to the approved goals.

Page intentionally left blank

DOMAIN 2—STRATEGIC MANAGEMENT (20%)

GS2-1 Which of the following drivers will **BEST** justify the decision of the board of directors of an enterprise to modify the IT strategy?

A. An update of a standard that addresses the alignment of the business and IT
B. A significant change in the enterprise architecture (EA)
C. A decision to merge the IT department with the finance department
D. A reduction in total cost of ownership (TCO) of IT projects

B is the correct answer.

Justification:
A. Standards support the implementation of the IT strategy.
B. The enterprise architecture (EA) is an integral part of the IT strategy; therefore, this would justify a need to modify the IT strategy.
C. Organizational changes are put in place to support the implementation of the IT strategy.
D. Financial implications are included in the IT strategy.

GS2-2 Which of the following choices **BEST** supports senior management in making decisions on IT project priorities?

A. An independent technical expert
B. Use of IT with market leaders
C. Use of IT by competitors
D. Well-documented business cases

D is the correct answer.

Justification:
A. An independent technical expert may support management in understanding technology; however, decisions remain based on business cases.
B. Aligning use of IT with market leaders may not help. Use of IT is best aligned with business objectives.
C. Understanding competitors' IT usage may be helpful in developing business cases.
D. Decisions on IT initiatives are based on business cases, which combine the business contribution and the risk profile of the projects.

GS2-3 An enterprise is planning to purchase a cloud service provider business that can only be supported by the use of unproven technology. Which of the following choices should be the **FIRST** to be considered?

A. Reprioritization of current work
B. Registration to an approved technical service provider list
C. Assessment of the balance between risk and opportunity
D. A plan for disaster recovery

C is the correct answer.

Justification:
A. Engaging in the purchase of the cloud business may lead to the reprioritization of current work.
B. A cloud service provider must be registered to an approved technical service provider list. This should be completed as part of the due diligence process.
C. When a new business is being evaluated, striking the balance between risk and opportunity is the fundamental approach from a governance perspective. Governance practitioners should make sure that the business opportunity is within the acceptable risk.
D. A strategic plan for disaster recovery should be completed as part of the due diligence process.

GS2-4   Which of the following choices is the **MOST** appropriate to provide senior management with the status of ongoing business?

    A.   A dashboard
    B.   A status meeting
    C.   Benchmark testing
    D.   A quality assurance (QA) report

A is the correct answer.

Justification:
    A.   A dashboard provides a structured report of the status of the current IT initiatives and highlights the necessary course of action for the required senior officer.
    B.   A dashboard is required for the status meeting.
    C.   Benchmarking is a high-level comparison of current status against industry standards and may be included in a dashboard.
    D.   Quality assurance (QA) reports also can be part of the dashboard.

GS2-5   When there is an anticipated change in the existing enterprise architecture (EA), which of the following choices is the **BEST** to consider to ensure alignment with the strategic planning process?

    A.   Adjustment of the control framework
    B.   Integration of the initiative in the IT portfolio
    C.   Establishment of change management accountability
    D.   Reprioritization of IT resource utilization

B is the correct answer.

Justification:
    A.   Adjustment of the control framework may be necessary; however, a control framework is not an enabler of alignment.
    B.   To guarantee continued alignment with business objectives when a change in enterprise architecture (EA) is involved, the initiatives must be integrated into the IT portfolio.
    C.   Establishment of change management accountability is necessary to guarantee execution of the initiative, but does not guarantee alignment.
    D.   Reprioritization of IT resource utilization is necessary to guarantee execution of the initiative, but does not guarantee alignment.

GS2-6 Which of the following choices **BEST** provides assurance to senior management on effective investment in IT?

    A.   Realization of the expected benefits for initiatives
    B.   The number of projects completed on time, quality and budget
    C.   Ongoing alignment of IT strategies with business strategies
    D.   An effective IT program and portfolio management

C is the correct answer.

Justification:
    A.   Realizing expected benefits for initiatives can be the result of the alignment of business and IT strategies.
    B.   The number of projects completed on time, quality and budget is a key performance indicator (KPI) of IT performance based on the alignment with business strategies.
    C.   Business strategies drive initiatives, and IT strategies support the business strategies, thus assuring senior management on effective investment in IT.
    D.   An effective IT program and portfolio management is a process for supporting the alignment of business and IT strategies.

GS2-7 Which of the following outputs is the **MOST** important for developing an enterprise architecture (EA) vision?

    A.   A strategic road map
    B.   Enterprise strategy
    C.   Organizational structure
    D.   A value proposition

D is the correct answer.

Justification:
    A.   A strategic road map is a document that identifies the steps to be taken by an enterprise to accomplish a strategic goal (e.g., sell online), describes why the goal is in place (e.g., compete more effectively) and how it will do it (e.g., build e-commerce web site). The enterprise architecture (EA) vision uses the enterprise strategic road map to propose a solution (e.g., architecture defining business processes, information, data, application and technology) that will help the enterprise achieve the strategic goal.
    B.   Enterprise strategy is the business strategy, formulated by the board of directors and senior management. The strategy is considered by all enterprise initiatives, including the development of an EA vision.
    C.   Organizational structure defines the groupings and ranking of people by business sector, geographic location, business function, level of authority and reporting hierarchy. It is typically considered when developing the EA vision.
    D.   ISACA's COBIT 5 framework states, "The architecture vision provides a first-cut, high-level description of the baseline and target architectures, covering the business, information, data, application and technology domains. The architecture vision provides the sponsor with a key tool to sell the benefits of the proposed capability to stakeholders within the enterprise. The architecture vision describes how the new capability will meet enterprise goals and strategic objectives and address stakeholder concerns when implemented."

GS2-8   In order for an IT policy to be most effective, it is **BEST** to ensure which of the following choices?

    A.   Consistency with enterprise-level policies
    B.   Approval by IT management
    C.   Dissemination through an awareness program
    D.   Review periodically for updates

A is the correct answer.

Justification:
    A.   **Consistency of IT policies at the enterprise level is a necessary step toward enterprise alignment.**
    B.   IT policies are owned by IT management. However, IT management ownership, without being consistent with corporate policies, may not further the enterprise goals.
    C.   Each employee is responsible for adherence to the relevant policies. However, adherence to policies that are not consistent with corporate policies will not further enterprise goals.
    D.   All policies should be periodically reviewed for continued applicability and consistency with enterprise policies.

GS2-9   Which of the following choices **MOST** inhibits enterprise architecture (EA) effectiveness? When EA is:

    A.   business owned.
    B.   derived from an industry framework.
    C.   an IT design document.
    D.   a strategic enabler.

C is the correct answer.

Justification:
    A.   An enterprise architecture (EA) should be business owned and IT supported to ensure that the four dimensions of EA are considered.
    B.   Implementation of the EA can be facilitated by the use of existing industry frameworks.
    C.   **EA encompasses business, applications, data and infrastructure. By considering an IT design document, the business aspects of the architecture are not fully considered.**
    D.   The EA is a strategic enabler because it considers the "to be" state of the enterprise.

GS2-10   Which of the following steps is the **MOST** important for developing an IT strategic plan?

    A.   Establish needs and priorities for information systems (IS) applications.
    B.   Determine how IT can extend the business strategy.
    C.   Determine the digital strategy to increase customer intimacy.
    D.   Determine how the company can differentiate by using new technologies.

B is the correct answer.

Justification:
    A.   Establishing needs and priorities for information systems (IS) applications is an important element of IT governance, but it can only be effective when priorities setting is based on the right IT strategies, which must be aligned with the business strategy.
    B.   **Determining how IT can extend the business strategy is the most important step for developing an IT strategic plan because IT strategy can support or extend the business strategy. Determining how IT can extend the business strategy is one of the roles of an IT strategy. This is necessary to ensure that IT management has completely understood the business strategies and has been able to translate them into IT strategies. If this step is not in place, there is a risk that the IT department will not focus on the business objectives and will, therefore, not deliver the expected capabilities is extremely high.**
    C.   Determining the digital strategy to increase customer intimacy is one possible direction for the company, but the first step is to determine in which direction the company wants to go.
    D.   Use of new technologies can help in the implementation of the company strategy when the direction has been determined by the board of directors. IT can extend the business strategy through the use of technology, but this is only one possibility.

GS2-11   Which of the following enablers is the **MOST** effective for the IT strategic planning process?

    A.   Focus on organizational effectiveness.
    B.   Develop a financial plan and business cases.
    C.   Choose the markets to serve.
    D.   Consider all options.

D is the correct answer.

Justification:
    A.   Organizational or operational effectiveness is the way that the enterprise must operate in order to deliver the necessary performance. A focus on organizational effectiveness is one aspect of the strategic planning analysis.
    B.   Developing a financial plan and business cases is an important element of defining the implications of the agreed-on strategies defined during the strategic planning process. This is the responsibility of operational management and finance department.
    C.   Choosing the markets to serve can contribute to the strategy, but is not the enabler; it is an implication of the defined strategy.
    D.   **Considering all options is the most effective enabler of the strategic planning process because it allows management to think "outside of the box" and find new or different ways to operate the business. Analyzing how rivals, competitors and new entrants in the field are currently running their businesses is a good way to achieve the objectives.**

GS2-12   Which of the following roles is the **MOST** influential one that the board of directors must play in the governance of enterprise IT?

    A.   Asking questions to challenge and focus on priorities
    B.   Providing an endorsement for large technology investments
    C.   Serving as a mechanism to reduce company politics
    D.   Helping IT to identify outsourcing opportunities

A is the correct answer.

Justification:
    A.   **Because the board of directors is composed of executive and nonexecutive directors, the board members can have a different perspective than executives inside the organization due to their oversight roles. By asking questions and focusing on priorities, the organization is forced to identify the most important capabilities to deliver in order to contribute to the business strategy.**
    B.   Providing an endorsement for large technology investments is also a role that the board of directors can play, but is secondary to providing focus on priorities.
    C.   When the board of directors focuses on priorities for decision making, this will provide clarity and will reduce company politics.
    D.   The role of the board of directors is to determine the sourcing principles (i.e., focus on core activities and noncore activities that have been outsourced), while IT management's responsibility is to identify outsourcing candidates in line with the principles defined by the board of directors.

DOMAIN 3—BENEFITS REALIZATION (16%)

GS3-1   Measuring the value of new services provided by IT to the business **BEST** helps an enterprise with:

A.   optimizing the risk associated with use of IT.
B.   implementing an IT governance framework.
C.   optimizing the portfolio of IT investments.
D.   maximizing stakeholder satisfaction.

C is the correct answer.

Justification:
A.   Managing risk will influence the way value is generated.
B.   Implementing an IT governance framework will contribute to the value of new services. Having the proper IT portfolio process optimized has a more holistic impact.
C.   **IT supports business processes in delivering services. To determine the prioritization of IT investments and manage portfolios that measure the value of IT services provides the best help.**
D.   Measuring value does not create stakeholder satisfaction, but delivering value will create stakeholder satisfaction.

GS3-2   Which of the following choices **BEST** helps in prioritizing IT improvement initiatives?

A.   Earned value analysis (EVA)
B.   Expected benefits realization
C.   Budget variance analysis
D.   A balanced scorecard (BSC)

B is the correct answer.

Justification:
A.   Earned value analysis (EVA) consists of comparing the following metrics at regular intervals during the project: budget to date, actual spending to date and estimate to complete and estimate at completion. EVA does not compare or analyze business outcomes. This is comparatively less effective than enterprise architecture (EA).
B.   **Prioritization of IT improvement initiatives based on business benefits and risk will provide the greatest benefits in total, and will help in prioritizing projects within a project portfolio.**
C.   Budget variance analysis compares budgeted and actual figures. This is not beneficial until after the project has been completed to determine whether the project met or exceeded the projected budget. This does not help in prioritizing IT initiatives.
D.   The balanced scorecard (BSC) is a performance reporting tool of IT. The BSC can be used to measure where an enterprise is both before and after the initiative has been put into place.

GS3-3 Which of the following choices is the **BEST** to present a business benefits realization initiative to the board of directors?

    A.   Business manager
    B.   Chief information officer (CIO)
    C.   Chief risk officer (CRO)
    D.   Chief financial officer (CFO)

A is the correct answer.

Justification:
    A.   **The business manager benefiting from the initiative needs to present the initiative and defend his/her case.**
    B.   The chief information officer (CIO) would support the business manager in presenting and defending his/her case, and, in specific cases, may be the business manager for IT projects.
    C.   The chief risk officer (CRO) would support the business manager in presenting and defending his/her case, and, in specific cases, may be the manager for risk-driven projects.
    D.   The chief financial officer (CFO) would support the business manager in presenting and defending his/her case, and, in specific cases, may be the manager for financial projects.

GS3-4 Which of the following choices is **MOST** essential to the overall success of an IT governance implementation?

    A.   Monitoring individual IT project costs
    B.   Formalizing monthly meetings between the board of directors and the chief information officer (CIO)
    C.   Monitoring key drivers for IT initiatives
    D.   Allocating required resources for individual projects

C is the correct answer.

Justification:
    A.   Monitoring individual project costs is a subset of monitoring key drivers for IT initiatives.
    B.   While the chief information officer (CIO) needs to provide relevant information to the board of directors, formal monthly meetings are not the only means of effective communication.
    C.   **Each IT initiative is related to a business objective contribution, and monitoring key delivery drivers on an ongoing basis provides the data to determine success.**
    D.   One of the key drivers for IT initiatives is the allocation of required resources for individual projects.

GS3-5 Which of the following activities would **BEST** motivate stakeholders to deliver value from major IT initiatives?

    A.   Incentivizing new ideas with new technology tools
    B.   Building accountability into the business case
    C.   Building ownership into the business case
    D.   Implementing objective setting and appraisal mechanisms

C is the correct answer.

Justification:
    A.   Incentivizing new ideas with new technology tools can be effective; however, it would not serve as the best motivator.
    B.   Building accountability clarifies the role, but ownership empowers the stakeholder to make decisions.
    C.   **When stakeholder ownership is built into the business case, everyone is clear about who owns the initiative during the life cycle of the IT initiative. This empowerment and role clarity motivates the stakeholder to deliver value.**
    D.   Implementing objective setting and appraisal mechanisms are an important part of the business case; however, this only creates a framework and does not provide motivation.

GS3-6 An enterprise does not currently have an online ordering process and is about to approve a business case for implementing an online ordering process. Which of the following choices should be the **PRIMARY** consideration before approving the initiative?

A. Expected realization is beneficial.
B. Total exposure is within risk tolerance.
C. Competitors are considering similar initiatives.
D. The enterprise expects to explore new markets.

A is the correct answer.

Justification:
A. **New initiative approvals are subject to a combination of benefits realization and risk mitigation.**
B. Risk tolerance is just one parameter for expected realization.
C. A competitor's initiatives are one parameter for expected realization.
D. Possible growth by exploring new markets is part of benefits realization.

GS3-7 Optimal value is generated by an IT-enabled investment portfolio through effective value management practices in the enterprise. Which of the following metrics **BEST** provides information about the performance of these management practices?

A. The percent of IT value drivers mapped to business value drivers
B. The percent of defined and approved business cases for the overall portfolio
C. The satisfaction level of key stakeholders regarding the accuracy of IT financial information
D. The percent of IT-enabled investments in the overall portfolio managed through the full life cycle

D is the correct answer.

Justification:
A. The percent of IT value drivers mapped to business value drivers is one component of the value management process. It provides information about the alignment of IT and business strategy, but not about getting value through the full effective value management practices.
B. The percent of defined and approved business cases for the overall portfolio measures the effectiveness of the decisions being made, not the value being delivered.
C. The accuracy of IT financial information is the responsibility of the finance area. This will not give any indication of how the practices deliver the necessary value.
D. **Effective value management practices include developing a business case, monitoring and managing progress, and evaluating results as described by the IT governance framework. The percent of IT-enabled investments in the overall portfolio managed over the full life cycle is the best indicator to measure the effectiveness of these practices.**

GS3-8   Which of the following performance indicators is the most **IMPORTANT** to ensure benefits realization?

    A.   The number of programs and projects needing rework due to quality
    B.   The number of programs and projects delivered on time and within budget
    C.   The number of identified types of risk of IT-enabled investments
    D.   The number of business cases being documented

B is the correct answer.

Justification:

    A.   The number of programs and projects needing rework due to quality is a good indicator of the quality of solutions being delivered, but this is not the most important performance indicator.
    B.   The number of programs and projects delivered on time and within budget is the most important performance indicator for benefits realization because two of the three key indicators of benefits realization would be achieved. The third key indicator of benefits realization is meeting business objectives.
    C.   The number of identified types of risk of IT-enabled investments provides information about what can go wrong, but does not provide any information about what does go wrong. Therefore, it is just an indicator and does not provide any information on benefits realization.
    D.   The number of business cases being documented is a necessary, but not sufficient, step to ensure benefits realization.

GS3-9   Which of the following IT governance practices is part of portfolio management?

    A.   Translating the strategic direction into a target investment mix
    B.   Developing and evaluating the initial program concept business case
    C.   Establishing informed and committed leadership
    D.   Defining portfolio characteristics

A is the correct answer.

Justification:

    A.   Translating the strategic direction into a target investment mix is the first activity within portfolio management. It ensures the clarity of the enterprise and IT strategies. The right target mix will achieve an optimal balance between long- and short-term returns, high and low risk, and financial and nonfinancial benefits.
    B.   Developing and evaluating the initial program concept business case is a typical investment management practice. A program business case will be submitted to the portfolio management process for prioritization.
    C.   Establishing informed and committed leadership needs to be addressed to ensure that value management practices are embedded in the enterprise. It is a prerequisite of an effective portfolio management capability.
    D.   Defining the portfolio characteristics is a governance activity needed to make the portfolio management process operational.

GS3-10   Which of the following choices is the **MAJOR** benefit of preparing a business case for a system development project?

      A.   Mitigation of IT risk involved in the project
      B.   Identification of relevant business stakeholders
      C.   Elimination of contending solution options
      D.   Confirmation of management sponsorship

D is the correct answer.

Justification:
A.   Strategic risk is included in the business case. Mitigation of IT risk is an operational activity.
B.   Identification of relevant business stakeholders is an important element of the business case, but it is of no benefit without buy-in from management.
C.   The business case will identify solution options with a recommendation for the most appropriate option.
D.   **A business case documents the rationale for making a business investment, used both to support a business decision on whether to proceed with the investment and as an operational tool to support management of the investment through its full economic life cycle. The business case is used to obtain management buy-in, thereby accepting accountability for benefits realization.**

Page intentionally left blank

CGEIT Review Questions, Answers & Explanations Manual 2015 Supplement

DOMAIN 4—RISK OPTIMIZATION (24%)

GS4-1 Which of the following choices is the **PRIMARY** role of IT senior management in risk management?

A. Reporting the results of IT-related risk to the enterprise risk portfolio
B. Identifying the enterprise appetite for IT risk management
C. Identifying the business impact of IT risk
D. Understanding IT risk management

A is the correct answer.

Justification:
A. **The role of IT senior management in risk management is to identify and report all IT-related risk to the enterprise risk portfolio.**
B. Identifying the enterprise appetite for risk management is the responsibility of the board of directors **and** senior management.
C. Identifying the business impact of IT risk is the responsibility of business management and business process owners.
D. Understanding risk management is the responsibility of each employee.

GS4-2 Which of the following criteria is the **MOST** important for selecting a key risk indicator (KRI)?

A. Relevance
B. Quantifiability
C. Sensitivity
D. Reliability

A is the correct answer.

Justification:
A. **Unless the key risk indicator (KRI) is relevant to the enterprise, other criteria do not matter.**
B. If the KRI is quantifiable, but not relevant to the enterprise, it cannot be used.
C. The sensitivity criterion describes the required threshold to trigger the KRI.
D. If the KRI measurement is reliable, but not relevant to the enterprise, it cannot be used. The reliability criterion describes whether the KRI flags an exception every time it occurs.

GS4-3 Which of the following approaches is the **BEST** to identify IT risk?

A. Ensuring that each IT risk has an associated mitigation
B. Ensuring that each IT risk aligns with legal and regulatory requirements
C. Ensuring that each IT risk maps to a business risk
D. Ensuring that each IT risk has buy-in from senior management

C is the correct answer.

Justification:
A. Mitigation is the operational response to the identified IT risk.
B. Some, but not all, risk may align with legal and regulatory requirements.
C. **Because IT is an enabler of at least one business process to support strategic objectives, every IT risk maps to a business risk.**
D. Buy-in from senior management ensures management oversight and direction for the identification and mitigation of the IT risk.

GS4-4 Which of the following choices is the **MAIN** result of having IT involved with business continuity planning?

    A.   To ensure IT system resilience
    B.   To ensure the realization of the business resilience objectives
    C.   To ensure that IT disaster recovery mechanisms are implemented
    D.   To ensure that IT drives the business continuity initiatives for the enterprise

B is the correct answer.

Justification:
    A.   IT system resilience is a subset of the business continuity plan.
    B.   Involving IT in business continuity planning will ensure that the IT-related risk is also covered.
    C.   IT disaster recovery mechanisms are a subset of IT system resilience, which is included in the business continuity plan (BCP).
    D.   Senior management, rather than IT, owns the business continuity initiative.

GS4-5 In the context of creating an IT risk management approach to business resiliency, the **GREATEST** advantage of performing a business impact analysis (BIA) is:

    A.   having an updated register reflecting all asset changes.
    B.   having both qualitative and quantitative estimations of risk.
    C.   eliminating the need to perform a risk analysis.
    D.   creating awareness about the business impact of losing critical IT assets.

D is the correct answer.

Justification:
    A.   A business impact analysis (BIA) will need to be updated periodically to reflect any asset changes.
    B.   A BIA should utilize both qualitative and quantitative estimates.
    C.   A BIA does not eliminate the need to perform a risk analysis.
    D.   A BIA enables raising the level of awareness for business continuity within the enterprise about business impact when critical IT assets are lost.

GS4-6 To create an effective IT risk management process, enterprises need to upgrade their risk dialogue between the board of directors and senior management. Which of the following choices would **BEST** initiate an effective communication process?

   A. Establishing regular meetings between IT management and the audit committee
   B. Developing a regular risk report for the board of directors on the progress of each IT risk mitigation
   C. Defining the risk roles for the board of directors and IT governance committees
   D. Implementing an IT risk training program for the board of directors

C is the correct answer.

Justification:
   A. The responsibility for regular meetings between IT management and the audit committee is included in defining the risk roles for the board of directors and IT governance committees.
   B. Reporting on the performance of measures that address key risk is included in defining the risk roles for the board of directors and IT governance committees.
   C. **Defining the risk roles for the board of directors and IT governance committees will start with a clear understanding of roles and risk responsibilities. By allocating responsibilities clearly, the board of directors and senior management can begin to explore the different aspects of IT risk that may impact the enterprise and to develop a plan to address and measure the effectiveness of their response.**
   D. Providing the board of directors with the necessary tools, including IT risk training, will help the board of directors to better understand IT-related risk and, therefore, will enable them to play the necessary role in IT risk management.

GS4-7 Which of the following choices **BEST** enables IT management and the board of directors to respond to regulatory requirements related to IT risk associated with outsourcing agreements?

   A. Establish IT risk management for third-party relationships.
   B. Develop an IT risk strategy for selecting third parties.
   C. Develop IT risk processes for overseeing third parties.
   D. Conduct independent IT risk reviews to effectively manage third parties.

A is the correct answer.

Justification:
   A. **Establishing an IT risk management process for third-party relationships best enables IT management to respond to regulatory requirements because they will be able to manage the entire third-party relationship life cycle from identification to termination.**
   B. Developing an IT risk strategy for selecting third parties is a subset of the third-party IT risk management process.
   C. Developing IT risk processes for overseeing third parties is a subset of the third-party IT risk management process.
   D. Conducting independent IT risk reviews to effectively manage third parties is a subset of the third-party IT risk management process.

GS4-8   Which of the following choices **BEST** describes the goal of developing effective key risk indicators (KRIs)?

    A.   Identify underperforming indicators of an enterprise.
    B.   Provide insights about risk events that affect the enterprise.
    C.   Develop an effective reporting mechanism to the board.
    D.   Identify potential risk for the enterprise's objectives.

D is the correct answer.

Justification:
    A.   Identifying underperforming indicators of an enterprise refers to key performance indicators (KPIs), not key risk indicators (KRIs).
    B.   KRIs are forward-looking and are not related to past events. KPIs refer to events that have impacted the enterprise.
    C.   Developing an effective reporting mechanism to the board of directors can only be done when the potential risk to enterprise objectives have been identified and addressed.
    D.   Identifying potential risk for the enterprise's objectives is the main goal of KRIs. The goal is to identify those potential events that could hinder the enterprise's ability to achieve its control objectives and develop measures that eliminate or minimize their potential impact.

GS4-9   A consultant is hired to support the risk management process for an enterprise. The consultant identifies that the enterprise has not implemented a risk management framework. Which of the following choices is the **BEST** reason for implementing a risk management framework?

    A.   Reduce IT risk.
    B.   Facilitate communication.
    C.   Reduce enterprise costs.
    D.   Provide risk reporting.

B is the correct answer.

Justification:
    A.   The first step to reduce IT risk is to have a risk framework to promote good communication. However, risk will not be reduced without the necessary associated activities.
    B.   Having a risk management framework includes common language for communication for all stakeholders.
    C.   A risk framework and effective risk management will not necessarily reduce costs. It is possible that as risk is recognized, costs will increase.
    D.   Risk reporting can exist whether or not a risk management framework is in place. However, having risk reporting based on a risk management framework will provide more clarity.

DOMAIN 4—RISK OPTIMIZATION

GS4-10   What is the **MOST** important implementation risk associated with a business process reengineering project?

A.   Scope risk
B.   User support risk
C.   Leadership risk
D.   Cultural risk

C is the correct answer.

Justification:
A.   Serious problems will arise if the scope of the project is improperly defined. However, scope is defined during the design and, therefore, scope risk is a design risk.
B.   If users do not provide the necessary support for the business process reengineering project, the resulting processes may not be workable. User support is a part of the design phase as well as for testing in the implementation phase. However, the project will not even get to this phase without leadership support.
C.   **C-level executives may fail to provide enough support for the project to be successful.**
D.   Cultural risk is part of the design, implementation and operational phases of a project. It is management's job to provide mitigation for the cultural risk during all phases of the project.

GS4-11   A mid-sized company has identified risk with the use of IT. Due to the potential stakeholder benefits, the direction is to make maximum use of IT. Risk was identified and documented. After implementation of activities to mitigate IT risk, which of the following choices is then the remaining type of IT risk to the company?

A.   Residual risk
B.   Business risk
C.   Inherent risk
D.   Control risk

A is the correct answer.

Justification:
A.   **Residual risk is the remaining risk after management has implemented risk responses (i.e., mitigation). The residual risk is then reviewed to determine whether it is an acceptable business risk or whether additional risk mitigation activities should be implemented.**
B.   All risk is ultimately a business risk to the company and must be accepted as part of standard business practice.
C.   Inherent risk is the risk exposure before management activities are implemented to mitigate risk.
D.   Control risk is the risk that the control activities put into place may have residual risk.

GS4-12 IT risk associated with the outsourcing of IT services is **BEST** managed through the:

    A.   development of policies and procedures.
    B.   performance of due diligence audits.
    C.   creation of multiple sourcing strategies
    D.   inclusion of controls and service level agreements (SLAs) into contracts.

D is the correct answer.

Justification:
    A.   Policies and procedures are important, but are not the most effective way to manage third-party risk by themselves.
    B.   Audits and the right to audit are important, but this will occur after the fact or an event.
    C.   Creating multiple sourcing strategies is one of the enterprise's risk responses and is also important, but this will occur before a specific source is selected.
    D.   Mitigating controls and requirements is normally included in contracts and agreements.

GS4-13 Which of the following choices is **MOST** important when documenting IT risk related to conforming to legal requirements for financial management?

    A.   Enterprise reporting policies
    B.   Change management policies
    C.   Service delivery processes
    D.   Enterprise goals and objectives

D is the correct answer.

Justification:
    A.   Enterprise goals and objectives may result in enterprise reporting policies, which are tools to achieve the enterprise goals and objectives.
    B.   Change management policies implement legal requirements and are required controls for legal requirements. As such, they support the enterprise goals and objectives.
    C.   Service delivery processes are operational instructions to support enterprise goals and objectives.
    D.   Enterprise goals and objectives must exist before risk can be identified to affect their achievement. Enterprise goals and objectives generally include compliance with legal requirements. IT events and, thus, IT risk, affect the achievement of enterprise goals and objectives.

GS4-14 For business resiliency, a company has a business continuity plan (BCP) in place. The IT disaster recovery plan (DRP) includes ongoing IT backup and use of an offsite vendor data center. The IT DRP provides which of the following risk response options to an event?

A. Avoidance
B. Mitigation
C. Transference
D. Acceptance

B is the correct answer.

Justification:
A. The IT disaster recovery plan (DRP) provides activities if there is a disruption to be recovered. Avoidance of the disruption event takes place before the disruption actually occurs.
B. **The business continuity plan (BCP) is a process for business resiliency to withstand disruption. The IT DRP provides mitigation plans in the event of a disastrous disruption.**
C. Transference is when the responsibility for a risk event is transferred to another entity. In this example, the responsibility for the risk event is not transferred to the vendor.
D. Acceptance is when the risk event is accepted. For example, it may be determined that if a disruption occurs, the disruption will be expected to last less than six hours. In this case, the company may decide to accept the disruption and not initiate the disaster recovery process.

Page intentionally left blank

DOMAIN 5—RESOURCE OPTIMIZATION (15%)

GS5-1 In order to ensure that IT department employees work toward the achievement of enterprise goals, it is **MOST** effective to:

A. train staff to escalate concerns before they reach a critical condition.
B. incorporate strategically aligned objectives in performance appraisals.
C. set up dual reporting lines, functional and managerial.
D. incorporate personal development goals in performance appraisals.

B is the correct answer.

Justification:
A. An escalation program may support the achievement of business goals; however, its effectiveness may be limited.
B. **It is most effective to allow for performance appraisals to incorporate objectives aligned with business strategy. In this way, management can ensure that employees are striving toward the enterprise goals.**
C. Dual reporting lines may or may not help the achievement of business goals.
D. This approach will guarantee personal development, but will not guarantee alignment with enterprise goals.

GS5-2 Which of the following choices should be given the **GREATEST** attention by the board of directors when considering outsourcing data backups?

A. Vendor certification of its backup restore process
B. The impact of vendor access to confidential information
C. Reliability of the vendor's disaster recovery plan (DRP)
D. The turnaround time for recovery of production data

A is the correct answer.

Justification:
A. **Vendor certification will provide evidence of the use of good practices.**
B. Vendor certification is the first information the board of directors will request. Vendor access to confidential information could be a follow-up concern.
C. Vendor certification is the first information the board of directors will request. The reliability of the vendor's disaster recovery plan (DRP) could be a follow-up concern.
D. The turnaround time for data recovery should be included in the service level agreement (SLA).

GS5-3   Which of the following governance processes is the **MOST** essential for optimizing sourcing strategies for an enterprise?

A.   Defining service level agreements (SLAs)
B.   Surveying stakeholder satisfaction
C.   Reviewing financial reports
D.   Monitoring IT service delivery performance

D is the correct answer.

Justification:
A.   Defining service level agreements (SLAs) provides metrics to define IT service level expectations.
B.   Surveying stakeholder satisfaction provides one measurement of IT performance.
C.   Financial reports provide one measurement of IT performance.
D.   **In order to ensure that IT resource usage is optimal, management must be aware of the performance of IT resources and compare it with expected value delivery. The input for this activity is provided by service delivery measurement and monitoring process.**

GS5-4   The **INTIAL** objective of baselining current IT resource levels is to:

A.   allow reprioritization.
B.   measure performance.
C.   align resource planning with industry norms.
D.   assess cost-effectiveness of outsourcing contracts.

B is the correct answer.

Justification:
A.   Measuring current performance enables reprioritization of current IT resource levels and performance.
B.   **Measuring current performance for the baseline current IT resource levels is the initial step in optimizing resource usage.**
C.   Benchmarking with industry norms will provide additional information about current resource performance.
D.   Assessing cost-effectiveness of outsourcing provides information, but is not the initial objective.

GS5-5   When monitoring the performance of IT resources, which of the following choices is the **MOST** important in ensuring the achievement of enterprise objectives?

A.   New IT services are being delivered on time, budget and quality.
B.   IT services meet security and regulatory requirements.
C.   Current IT services perform according to service level agreement (SLA) expectations.
D.   IT supports business growth by technology-related innovation.

C is the correct answer.

Justification:
A.   New service delivery is definitely an important measurement of IT performance; however, current service delivery allows the enterprise to continue to operate.
B.   Security and regulatory requirement are part of the service level agreement (SLA) expectations.
C.   **Management must monitor the performance of IT resources to ensure that the expected benefits from IT are realized according to planned performance.**
D.   IT plays an important role in business growth through technology-related innovation; however, current service delivery allows the enterprise to continue to operate.

GS5-6   To ensure alignment with business strategy when using an external service provider/outsourcing vendor, it is **MOST** important to:

A.   design a knowledge transfer program.
B.   develop service level agreements (SLAs).
C.   agree on scope and service expectations.
D.   perform due diligence.

C is the correct answer.

Justification:
A.   Designing a knowledge transfer program is important, but scope and service expectations must be agreed on first.
B.   Scope and service expectations are agreed on before service level agreements (SLAs) are developed.
C.   **When external vendors are used, it is most important to agree on scope and service expectations to minimize the risk of misunderstanding of the contractual engagement. It is a fundamental step in outsourcing management.**
D.   Due diligence is performed prior to selecting an external vendor.

GS5-7   Which of the following choices is the **GREATEST** risk to IT governance if IT is outsourced?

A.   The enterprise may lose sight of its accountability of IT.
B.   Costs of running IT may escalate.
C.   The vendor may subcontract the job, leading to unsatisfactory results.
D.   In-house IT employees may lose critical experience.

A is the correct answer.

Justification:
A.   **Outsourcing IT may give management and the board of directors a false sense of nonaccountability for IT decisions.**
B.   Escalation of IT costs can be a result of the enterprise losing sight of its accountability of IT.
C.   Subcontracting is an operational issue to be blended into the outsourcing contract.
D.   The impact of the loss of critical experience by in-house IT employees may be mitigated by proper accountability of the enterprise with respect to IT.

GS5-8   Which of the following activities is the **MOST** important when developing a resource management plan for availability of IT resources?

A.   Assessing IT risk scenarios associated with resource unavailability
B.   Mapping IT resource supply with business demand
C.   Considering outsourcing options if skilled resources are not available
D.   Identifying resource unavailability against demand

B is the correct answer.

Justification:
A.   Assessing IT risk scenarios associated with resource unavailability is dependent on mapping IT resource supply with business demand.
B.   Mapping IT resource supply with business demand is the initial analysis activity for the resource management plan. This information is needed to understand the current situation.
C.   Considering outsourcing options if skilled resources are not available is a possible solution to resource unavailability based on mapping the IT resource supply.
D.   Identifying resource unavailability against demand is a subset of mapping IT resource supply with business demand.

GS5-9   In effective information governance, the board of directors needs to **FIRST**:

A.   define information attributes required for use by business functions.
B.   define the accountability and responsibility of the data privacy officer.
C.   establish a federated organizational model.
D.   define quality criteria for information across a range of different quality goals.

D is the correct answer.

Justification:
A.   Defining information attributes required for use by business functions is the responsibility of senior management.
B.   Defining the accountability and responsibility of the data privacy officer is the responsibility of senior management.
C.   Effective information governance can be achieved with different organizational models, not necessarily a federated one.
D.   Information should be treated as a corporate asset, like people, physical assets and goodwill. Accurate and trusted business information and insight can create value or be a liability for running operational processes and for decision making. Defining the quality criteria for information at the board level across a range of different quality goals includes relevancy, completeness and restricted access and may lead to effective information governance.

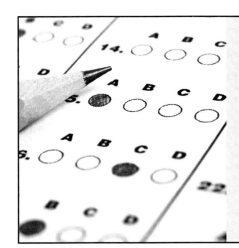

POSTTEST

If you wish to take a posttest to determine strengths and weaknesses, the Sample Exam begins on page 35 and the posttest answer sheet is on page 49. You can score your posttest with the Sample Exam Answer and Reference Key on page 45.

Page intentionally left blank

SAMPLE EXAM

1. Optimal value is generated by an IT-enabled investment portfolio through effective value management practices in the enterprise. Which of the following metrics **BEST** provides information about the performance of these management practices?

    A. The percent of IT value drivers mapped to business value drivers
    B. The percent of defined and approved business cases for the overall portfolio
    C. The satisfaction level of key stakeholders regarding the accuracy of IT financial information
    D. The percent of IT-enabled investments in the overall portfolio managed through the full life cycle

2. Which of the following choices is the **PRIMARY** role of IT senior management in risk management?

    A. Reporting the results of IT-related risk to the enterprise risk portfolio
    B. Identifying the enterprise appetite for IT risk management
    C. Identifying the business impact of IT risk
    D. Understanding IT risk management

3. Which of the following drivers will **BEST** justify the decision of the board of directors of an enterprise to modify the IT strategy?

    A. An update of a standard that addresses the alignment of the business and IT
    B. A significant change in the enterprise architecture (EA)
    C. A decision to merge the IT department with the finance department
    D. A reduction in total cost of ownership (TCO) of IT projects

4. Which of the following choices is the **MAIN** advantage of implementing a governance of enterprise IT framework?

    A. Establishing and monitoring accountability for IT-related initiatives
    B. Reducing IT-related risk by increasing IT investment
    C. Reducing IT-related costs by achieving IT process improvements
    D. Centralizing IT control through an IT steering committee

5. The **INTIAL** objective of baselining current IT resource levels is to:

    A. allow reprioritization.
    B. measure performance.
    C. align resource planning with industry norms.
    D. assess cost-effectiveness of outsourcing contracts.

6. Which of the following governance processes is the **MOST** essential for optimizing sourcing strategies for an enterprise?

    A. Defining service level agreements (SLAs)
    B. Surveying stakeholder satisfaction
    C. Reviewing financial reports
    D. Monitoring IT service delivery performance

7. In the context of creating an IT risk management approach to business resiliency, the **GREATEST** advantage of performing a business impact analysis (BIA) is:

    A. having an updated register reflecting all asset changes.
    B. having both qualitative and quantitative estimations of risk.
    C. eliminating the need to perform a risk analysis.
    D. creating awareness about the business impact of losing critical IT assets.

8.   When a new IT governance policy has been approved, it is **BEST** to:

    A.   have an independent party sign off.
    B.   conduct a walk-through exercise.
    C.   prepare a communication plan.
    D.   update the IT strategy accordingly.

9.   A consulting firm re-engineered a customer trading system of an investment bank. Then the investment bank requested a security review of this system from the same consulting firm. From an IT governance perspective, which of the following choices is the **BEST** to consider?

    A.   Ensure that sensitive customer data are securely kept inside the consulting firm.
    B.   Ensure that a security assurance review plan is in line with regulatory requirements.
    C.   Ensure that segregation of duties (SoD) is in place within the consulting firm.
    D.   Ensure that the service level meets the criteria in the vendor due diligence policy.

10.   Which of the following choices **MOST** inhibits enterprise architecture (EA) effectiveness? When EA is:

    A.   business owned.
    B.   derived from an industry framework.
    C.   an IT design document.
    D.   a strategic enabler.

11.   To create an effective IT risk management process, enterprises need to upgrade their risk dialogue between the board of directors and senior management. Which of the following choices would **BEST** initiate an effective communication process?

    A.   Establishing regular meetings between IT management and the audit committee
    B.   Developing a regular risk report for the board of directors on the progress of each IT risk mitigation
    C.   Defining the risk roles for the board of directors and IT governance committees
    D.   Implementing an IT risk training program for the board of directors

12.   In order to ensure that IT department employees work toward the achievement of enterprise goals, it is **MOST** effective to:

    A.   train staff to escalate concerns before they reach a critical condition.
    B.   incorporate strategically aligned objectives in performance appraisals.
    C.   set up dual reporting lines, functional and managerial.
    D.   incorporate personal development goals in performance appraisals.

13.   Which of the following choices is the **MOST** relevant in the enterprise culture change of an IT governance implementation?

    A.   Having employees who have values and beliefs
    B.   Having corporate values aligned with leaders in the industry
    C.   Having leaders who inspire new values
    D.   Having clearly communicated values and beliefs

14.   Which of the following activities is the **MOST** essential for ensuring resource optimization within governance of enterprise IT?

    A.   Providing direction for strategic resources
    B.   Defining guidelines for performance indicators
    C.   Evaluating resource strategy against enterprise requirements
    D.   Establishing principles for management of resources

15. Which of the following steps is the **MOST** important for developing an IT strategic plan?

   A. Establish needs and priorities for information systems (IS) applications.
   B. Determine how IT can extend the business strategy.
   C. Determine the digital strategy to increase customer intimacy.
   D. Determine how the company can differentiate by using new technologies.

16. An enterprise does not currently have an online ordering process and is about to approve a business case for implementing an online ordering process. Which of the following choices should be the **PRIMARY** consideration before approving the initiative?

   A. Expected realization is beneficial.
   B. Total exposure is within risk tolerance.
   C. Competitors are considering similar initiatives.
   D. The enterprise expects to explore new markets.

17. Which of the following performance indicators is the most **IMPORTANT** to ensure benefits realization?

   A. The number of programs and projects needing rework due to quality
   B. The number of programs and projects delivered on time and within budget
   C. The number of identified types of risk of IT-enabled investments
   D. The number of business cases being documented

18. When monitoring the performance of IT resources, which of the following choices is the **MOST** important in ensuring the achievement of enterprise objectives?

   A. New IT services are being delivered on time, budget and quality.
   B. IT services meet security and regulatory requirements.
   C. Current IT services perform according to service level agreement (SLA) expectations.
   D. IT supports business growth by technology-related innovation.

19. Which of the following outputs is the **MOST** important for developing an enterprise architecture (EA) vision?

   A. A strategic road map
   B. Enterprise strategy
   C. Organizational structure
   D. A value proposition

20. Which of the following choices is the **MAIN** result of having IT involved with business continuity planning?

   A. To ensure IT system resilience
   B. To ensure the realization of the business resilience objectives
   C. To ensure that IT disaster recovery mechanisms are implemented
   D. To ensure that IT drives the business continuity initiatives for the enterprise

21. Which of the following roles is the **MOST** influential one that the board of directors must play in the governance of enterprise IT?

   A. Asking questions to challenge and focus on priorities
   B. Providing an endorsement for large technology investments
   C. Serving as a mechanism to reduce company politics
   D. Helping IT to identify outsourcing opportunities

22. Which of the following choices **BEST** describes the role of the board of directors in IT risk governance?

    A.   Ensure that the planning, budgeting and performance of IT risk controls are appropriate.
    B.   Assess and incorporate the results of the IT risk management activity into the decision-making process.
    C.   Ensure that the enterprise risk appetite and tolerance are understood and communicated
    D.   Identify, evaluate and minimize risk to IT systems that support the mission of the enterprise.

23. Information security governance awareness is **BEST** established when:

    A.   senior management is supportive.
    B.   data ownership is identified.
    C.   assets to be protected are identified.
    D.   security certifications are issued.

24. Which of the following choices is the **MOST** appropriate to provide senior management with the status of ongoing business?

    A.   A dashboard
    B.   A status meeting
    C.   Benchmark testing
    D.   A quality assurance (QA) report

25. To ensure alignment with business strategy when using an external service provider/outsourcing vendor, it is **MOST** important to:

    A.   design a knowledge transfer program.
    B.   develop service level agreements (SLAs).
    C.   agree on scope and service expectations.
    D.   perform due diligence.

26. Which of the following choices is **MOST** important when documenting IT risk related to conforming to legal requirements for financial management?

    A.   Enterprise reporting policies
    B.   Change management policies
    C.   Service delivery processes
    D.   Enterprise goals and objectives

27. In order for an IT policy to be most effective, it is **BEST** to ensure which of the following choices?

    A.   Consistency with enterprise-level policies
    B.   Approval by IT management
    C.   Dissemination through an awareness program
    D.   Review periodically for updates

28. Which of the following choices **BEST** provides assurance to senior management on effective investment in IT?

    A.   Realization of the expected benefits for initiatives
    B.   The number of projects completed on time, quality and budget
    C.   Ongoing alignment of IT strategies with business strategies
    D.   An effective IT program and portfolio management

29.   Which of the following choices demonstrates the **GREATEST** influence of an IT governance framework for IT-related issues?

    A.   To settle differences of opinion among board of directors members
    B.   To resolve cross-departmental conflicts
    C.   To gain loyalty from key stakeholders
    D.   To investigate weaknesses within processes

30.   When implementing governance of enterprise IT, which of the following factors is the **MOST** critical for the success of the implementation?

    A.   Improving IT knowledge of the board of directors
    B.   Decision making on IT investments by the board of directors
    C.   Documenting the IT strategy
    D.   Identifying the enablers and establishing performance measures

31.   Which of the following choices is the **BEST** to present a business benefits realization initiative to the board of directors?

    A.   Business manager
    B.   Chief information officer (CIO)
    C.   Chief risk officer (CRO)
    D.   Chief financial officer (CFO)

32.   A consultant is hired to support the risk management process for an enterprise. The consultant identifies that the enterprise has not implemented a risk management framework. Which of the following choices is the **BEST** reason for implementing a risk management framework?

    A.   Reduce IT risk.
    B.   Facilitate communication.
    C.   Reduce enterprise costs.
    D.   Provide risk reporting.

33.   What is the **MOST** important implementation risk associated with a business process reengineering project?

    A.   Scope risk
    B.   User support risk
    C.   Leadership risk
    D.   Cultural risk

34.   Which of the following choices has the **GREATEST** impact on the selection of an IT governance framework?

    A.   Corporate culture
    B.   Data regulatory requirements
    C.   Skills and competencies
    D.   Current process maturity level

35.   While implementing IT governance within an enterprise, the **PRIMARY** focus must be on the objectives of:

    A.   the enterprise.
    B.   the stakeholders.
    C.   the business function.
    D.   IT management.

36. Which of the following criteria is the **MOST** important for selecting a key risk indicator (KRI)?

    A. Relevance
    B. Quantifiability
    C. Sensitivity
    D. Reliability

37. Which of the following choices **BEST** helps in prioritizing IT improvement initiatives?

    A. Earned value analysis (EVA)
    B. Expected benefits realization
    C. Budget variance analysis
    D. A balanced scorecard (BSC)

38. Which of the following IT governance practices is part of portfolio management?

    A. Translating the strategic direction into a target investment mix
    B. Developing and evaluating the initial program concept business case
    C. Establishing informed and committed leadership
    D. Defining portfolio characteristics

39. Measuring the value of new services provided by IT to the business **BEST** helps an enterprise with:

    A. optimizing the risk associated with use of IT.
    B. implementing an IT governance framework.
    C. optimizing the portfolio of IT investments.
    D. maximizing stakeholder satisfaction.

40. Which of the following choices **BEST** describes the goal of developing effective key risk indicators (KRIs)?

    A. Identify underperforming indicators of an enterprise.
    B. Provide insights about risk events that affect the enterprise.
    C. Develop an effective reporting mechanism to the board.
    D. Identify potential risk for the enterprise's objectives.

41. Which of the following choices is the **MAJOR** benefit of preparing a business case for a system development project?

    A. Mitigation of IT risk involved in the project
    B. Identification of relevant business stakeholders
    C. Elimination of contending solution options
    D. Confirmation of management sponsorship

42. Which of the following choices **BEST** enables IT management and the board of directors to respond to regulatory requirements related to IT risk associated with outsourcing agreements?

    A. Establish IT risk management for third-party relationships.
    B. Develop an IT risk strategy for selecting third parties.
    C. Develop IT risk processes for overseeing third parties.
    D. Conduct independent IT risk reviews to effectively manage third parties.

43.   Which of the following benefits is the **MOST** important for senior management to understand the value of governance of enterprise IT? It allows senior management to:

      A.   understand how the IT department works.
      B.   make key IT-related decisions.
      C.   optimize IT resource utilization.
      D.   evaluate business continuity provisions.

44.   Which of the following choices should be given the **GREATEST** attention by the board of directors when considering outsourcing data backups?

      A.   Vendor certification of its backup restore process
      B.   The impact of vendor access to confidential information
      C.   Reliability of the vendor's disaster recovery plan (DRP)
      D.   The turnaround time for recovery of production data

45.   In effective information governance, the board of directors needs to **FIRST**:

      A.   define information attributes required for use by business functions.
      B.   define the accountability and responsibility of the data privacy officer.
      C.   establish a federated organizational model.
      D.   define quality criteria for information across a range of different quality goals.

46.   Which of the following choices is the **GREATEST** risk to IT governance if IT is outsourced?

      A.   The enterprise may lose sight of its accountability of IT.
      B.   Costs of running IT may escalate.
      C.   The vendor may subcontract the job, leading to unsatisfactory results.
      D.   In-house IT employees may lose critical experience.

47.   Which of the following enablers is the **MOST** effective for the IT strategic planning process?

      A.   Focus on organizational effectiveness.
      B.   Develop a financial plan and business cases.
      C.   Choose the markets to serve.
      D.   Consider all options.

48.   Which of the following approaches is the **BEST** to identify IT risk?

      A.   Ensuring that each IT risk has an associated mitigation
      B.   Ensuring that each IT risk aligns with legal and regulatory requirements
      C.   Ensuring that each IT risk maps to a business risk
      D.   Ensuring that each IT risk has buy-in from senior management

49.   An enterprise is planning to purchase a cloud service provider business that can only be supported by the use of unproven technology. Which of the following choices should be the **FIRST** to be considered?

      A.   Reprioritization of current work
      B.   Registration to an approved technical service provider list
      C.   Assessment of the balance between risk and opportunity
      D.   A plan for disaster recovery

50.   The **PRIMARY** focus in effective organizational change enablement of a governance of enterprise IT
        implementation should be on:

        A.   documenting the what and how of the change.
        B.   clarifying the reason to change.
        C.   communication of the vision.
        D.   demonstrating achieved results.

51.   When there is an anticipated change in the existing enterprise architecture (EA), which of the following
        choices is the **BEST** to consider to ensure alignment with the strategic planning process?

        A.   Adjustment of the control framework
        B.   Integration of the initiative in the IT portfolio
        C.   Establishment of change management accountability
        D.   Reprioritization of IT resource utilization

52.   Which of the following activities is the **MOST** important when developing a resource management plan for
        availability of IT resources?

        A.   Assessing IT risk scenarios associated with resource unavailability
        B.   Mapping IT resource supply with business demand
        C.   Considering outsourcing options if skilled resources are not available
        D.   Identifying resource unavailability against demand

53.   Which of the following activities would **BEST** motivate stakeholders to deliver value from major IT initiatives?

        A.   Incentivizing new ideas with new technology tools
        B.   Building accountability into the business case
        C.   Building ownership into the business case
        D.   Implementing objective setting and appraisal mechanisms

54.   Which of the following choices **BEST** supports senior management in making decisions on IT project priorities?

        A.   An independent technical expert
        B.   Use of IT with market leaders
        C.   Use of IT by competitors
        D.   Well-documented business cases

55.   For business resiliency, a company has a business continuity plan (BCP) in place. The IT disaster recovery
        plan (DRP) includes ongoing IT backup and use of an offsite vendor data center. The IT DRP provides
        which of the following risk response options to an event?

        A.   Avoidance
        B.   Mitigation
        C.   Transference
        D.   Acceptance

56.   Which of the following choices is the **PRIMARY** reason for defining and managing the enterprise
        IT strategy?

        A.   It has become an industry standard.
        B.   It directs short-term IT goals.
        C.   It improves the efficiency of IT services.
        D.   It contributes to business value.

57. A mid-sized company has identified risk with the use of IT. Due to the potential stakeholder benefits, the direction is to make maximum use of IT. Risk was identified and documented. After implementation of activities to mitigate IT risk, which of the following choices is then the remaining type of IT risk to the company?

    A. Residual risk
    B. Business risk
    C. Inherent risk
    D. Control risk

58. Which of the following choices would **PRIMARILY** create transparency in the IT decision-making process as part of governance of enterprise IT?

    A. Stakeholder-approved roles, responsibilities, goals and metrics are communicated.
    B. The progress reporting process of service delivery is clearly established.
    C. Communication of decisions to IT employees is clear.
    D. Balanced scorecard (BSC) results are promptly communicated to the enterprise.

59. Which of the following choices is **MOST** essential to the overall success of an IT governance implementation?

    A. Monitoring individual IT project costs
    B. Formalizing monthly meetings between the board of directors and the chief information officer (CIO)
    C. Monitoring key drivers for IT initiatives
    D. Allocating required resources for individual projects

60. IT risk associated with the outsourcing of IT services is **BEST** managed through the:

    A. development of policies and procedures.
    B. performance of due diligence audits.
    C. creation of multiple sourcing strategies.
    D. inclusion of controls and service level agreements (SLAs) into contracts.

Page intentionally left blank

CGEIT® Review Questions, Answers & Explanations Manual 2015 Supplement

SAMPLE EXAM ANSWER AND REFERENCE KEY

Exam Question #	Key	Ref. #	Exam Question #	Key	Ref. #
1	D	GS3-7	31	A	GS3-3
2	A	GS4-1	32	B	GS4-9
3	B	GS2-1	33	C	GS4-10
4	A	GS1-1	34	A	GS1-8
5	B	GS5-4	35	B	GS1-10
6	D	GS5-3	36	A	GS4-2
7	D	GS4-5	37	B	GS3-2
8	C	GS1-2	38	A	GS3-9
9	C	GS1-5	39	C	GS3-1
10	C	GS2-9	40	D	GS4-8
11	C	GS4-6	41	D	GS3-10
12	B	GS5-1	42	A	GS4-7
13	C	GS1-12	43	B	GS1-6
14	D	GS1-7	44	A	GS5-2
15	B	GS2-10	45	D	GS5-9
16	A	GS3-6	46	A	GS5-7
17	B	GS3-8	47	D	GS2-11
18	C	GS5-5	48	C	GS4-3
19	D	GS2-7	49	C	GS2-3
20	B	GS4-4	50	B	GS1-11
21	A	GS2-12	51	B	GS2-5
22	C	GS1-13	52	B	GS5-8
23	A	GS1-4	53	C	GS3-5
24	A	GS2-4	54	D	GS2-2
25	C	GS5-6	55	B	GS4-14
26	D	GS4-13	56	D	GS1-3
27	A	GS2-8	57	A	GS4-11
28	C	GS2-6	58	A	GS1-15
29	B	GS1-14	59	C	GS3-4
30	D	GS1-9	60	D	GS4-12

Reference example: GS5-6 = See domain 5, question 6, for explanation of the answer.

Page intentionally left blank

CGEIT® Review Questions, Answers & Explanations Manual 2015 Supplement

SAMPLE EXAM ANSWER SHEET (PRETEST)

Please use this answer sheet to take the sample exam as a pretest to determine strengths and weaknesses. The answer key/reference grid is on page 45.

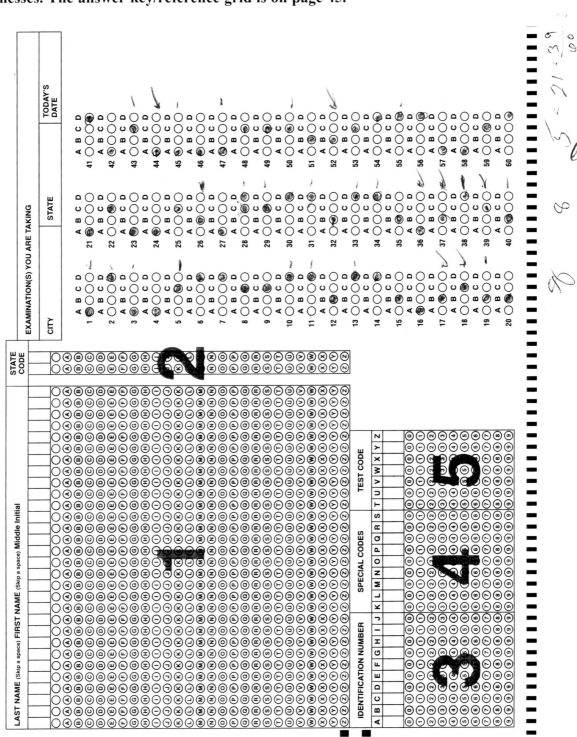

Page intentionally left blank

CGEIT® Review Questions, Answers & Explanations Manual 2015 Supplement
SAMPLE EXAM ANSWER SHEET (POSTTEST)

Please use this answer sheet to take the sample exam as a posttest to determine strengths and weaknesses. The answer key/reference grid is on page 45.

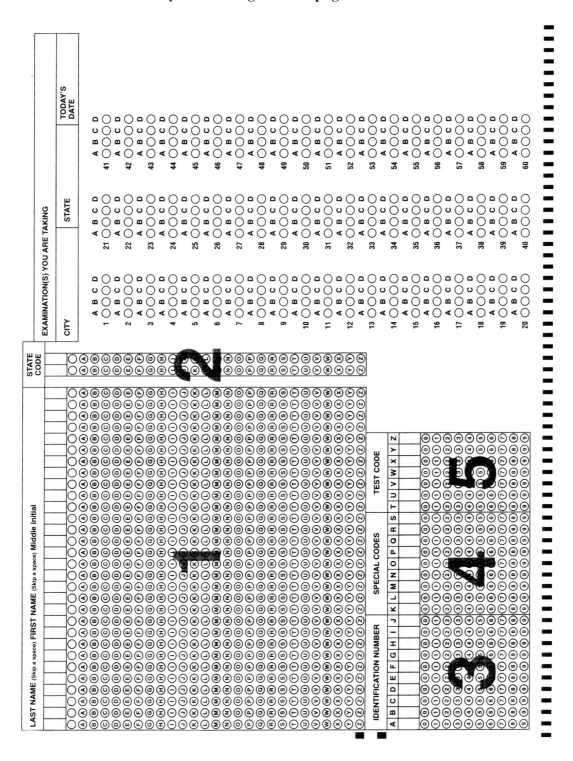

Page intentionally left blank

CGEIT Review Questions, Answers & Explanations Manual 2015 Supplement

EVALUATION

ISACA continuously monitors the swift and profound professional, technological and environmental advances affecting the IT governance professional. Recognizing these rapid advances, CGEIT review manuals are updated annually.

To assist ISACA in keeping abreast of these advances, please take a moment to evaluate the *CGEIT® Review Questions, Answers & Explanations Manual 2015 Supplement*. Such feedback is valuable to fully serve the profession and future CGEIT exam registrants.

To complete the evaluation, please visit *www.isaca.org/studyaidsevaluation*.

Thank you for your support and assistance.

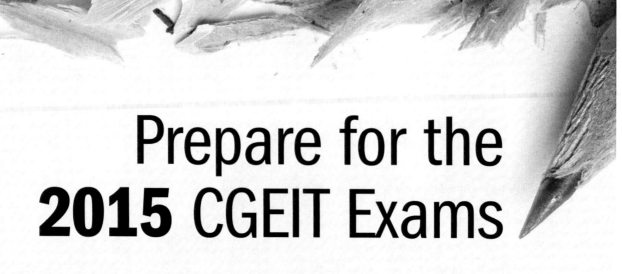

Prepare for the
2015 CGEIT Exams

2015 CGEIT Review Resources for Exam Preparation and Professional Development

Successful Certified in the Governance of Enterprise IT® (CGEIT®) exam candidates have an organized plan of study. To assist individuals with the development of a successful study plan, ISACA® offers several study aids and review courses to exam candidates. These include:

Study Aids

- *CGEIT® Review Manual 2015*

- *CGEIT® Review Questions, Answers & Explanations Manual 2015*

- *CGEIT® Review Questions, Answers & Explanations Manual 2015 Supplement*

To order, visit *www.isaca.org/cgeitbooks.*

Review Courses

- Chapter-sponsored review courses *(www.isaca.org/cgeitreview)*

Trust in, and value from, information systems